YOUNG FIRE EQUIPMENT
1932 THROUGH 1991
PHOTO ARCHIVE

Edited by Leo E. Duliba
& Tom W. Shand

Iconografix
Photo Archive Series

Iconografix
PO Box 446
Hudson, Wisconsin 54016 USA

Library of Congress Card Number: 00-132394

ISBN 1-58388-015-1

00 01 02 03 04 05 06 5 4 3 2 1

Printed in the United States of America

Cover and book design by Shawn Glidden

Copy Editing by Dylan Frautschi

Cover Photo Caption: West Islip, New York, 68-279. Young produced twin Crusader series pumpers for the West Islip Fire Department. The rescue style compartments were known as Kenilworth style bodies and featured six enclosed compartments with ground ladders, hard suction hose, and pike poles stored below the hose bed. This pumper, delivered to the Mohawk Engine Company, was equipped with a Hale 1500-gpm pump, 300-gallon water tank, and a special hose body to accommodate 3,450 feet of 3-inch hose, 400 feet of 2 1/2-inch hose, and 800 feet of 1 1/2-inch hose.

Book Proposals

Iconografix is a publishing company specializing in books for transportation enthusiasts. We publish in a number of different areas, including Automobiles, Auto Racing, Buses, Construction Equipment, Emergency Equipment, Farming Equipment, Railroads & Trucks. The Iconografix imprint is constantly growing and expanding into new subject areas.

Authors, editors, and knowledgeable enthusiasts in the field of transportation history are invited to contact the Editorial Department at Iconografix, Inc., PO Box 446, Hudson, WI 54016.

PREFACE

The histories of machines and mechanical gadgets are contained in the books, journals, correspondence, and personal papers stored in libraries and archives throughout the world. Written in tens of languages, covering thousands of subjects, the stories are recorded in millions of words.

Words are powerful. Yet, the impact of a single image, a photograph or an illustration, often relates more than dozens of pages of text. Fortunately, many of the libraries and archives that house the words also preserve the images.

In the *Photo Archive Series,* Iconografix reproduces photographs and illustrations selected from public and private collections. The images are chosen to tell a story—to capture the character of their subject. Reproduced as found, they are accompanied by the captions made available by the archive.

The Iconografix *Photo Archive Series* is dedicated to young and old alike, the enthusiast, the collector and anyone who, like us, is fascinated by "things" mechanical.

ACKNOWLEDGMENTS & DEDICATIONS

Without the help of Richard Young and his family, this book would not have been possible. Mister Young gave us full access to not only his corporate files, but turned over all of the archive photos in his former company's possession. The legacy that this family has handed down to the American Fire Apparatus Industry cannot go unrecognized. The Young family's involvement through the years left deep imprints in such companies as Buffalo, Case, Cayasler, Young, and R.D. Murray. From grandfather, Lester, to sons, Richard and Lester Jr., to grandson, Tim, the spirit has not died.

One other person cannot go without mention. Fire apparatus historian and writer, Joe Raymond, was the first to pursue the Young story. Without his vital knowledge and input into this book, it would not have been completed. Joe is indeed the third and missing editor of this book. Dick Young has called we three apparatus historians his last "YOUNG CRUSADERS"; Joe, Leo, and Tom take deep pride in that kind knighting.

Several apparatus historians and photographers have also provided valuable information to us. They include: Mark V. Carr, William Friedrich, Denis Gardner, Donald L. Loeb, Robert B. Milnes, Larry Phillips, Shaun P. Ryan, John Toomey, Glenn Vincent, and Joel Woods. Their assistance is gratefully acknowledged.

And finally to all apparatus chauffeurs and engineers who operate these units. Their dedication and skills enable fire fighters to perform all of the required tactics and arrive back to the station safely.

This photo shows the Buffalo, New York, William Street plant at its height. Young moved into the far-left portion of the building in 1936 and remained through four expansions until 1966. In that year, they moved to suburban Lancaster, New York. The Buffalo Fire Appliance Corporation plant was only two blocks to the left.

INTRODUCTION

In 1932 during the height of the Depression, Allen Case of Buffalo, New York, would create the infant beginnings of a fire apparatus manufacturing company in a small Ludington Street garage. From this humble start would grow the internationally known Young Fire Equipment Corporation.

The year 1933 saw the small company partnership of father and son, Allen and Richard Case, and friend, Nicholas Yavicoli, form a loosely related company called Casc Fire Equipment. Case would build a total of only five units before going through a name change the following year.

On January 22, 1934, two brothers would enter the picture. Theodore and Peter Geisler would join the other three partners to form a corporation. The name of this new company would consist of an anagram made up of parts of the various last names of the five partners. Cayasler Manufacturing Corporation would go on to produce quality fire apparatus until 1944.

In 1936, a new partner emerged in the form of Lester W. Young. Young was a vice president of Buffalo Fire Appliance Corporation, whose plants were located only a couple of city blocks down the street from Cayasler. Young would venture into his own manufacturing right when he left Buffalo Fire Appliance and purchase controlling interest in Cayasler. This move finally led to the "Y" in Cayasler, changing the company's corporate name to his own on April 5, 1944. From that date until 1991, the Young Fire Equipment Corporation would remain under Young family control.

In the early years of this company, most of the production of apparatus would center on small town needs in fire equipment. Through the Cayasler years, most pumpers were small, well-built front mounts with standard styling designs. In the later half of the 1930s, the company would sophisticate itself into more classic designs that were still geared toward small town usage.

During the early Young years, unique styling and innovations began to appear in the company's completed units. Many firsts in fire apparatus manufacturing started to appear and be noticed nationally. Among these concepts were the first use of metal ladder compliments on quad and city-service apparatus, first complete enclosure of ladder unit bodies, first true heavy rescue designs, double wall compartmentation, early use of Mattydale lay concept, and other milestone innovations. Following into and after the 1960s, many more such recognized Young standards would enter the competition nationally and propel this company into not only national but also international prominence and distribution.

Young began to utilize a sequential serial number system for apparatus in 1962. This format carried through until the last unit was delivered in 1991.

Leo E. Duliba

Ovid, New York, currently owns the oldest surviving example of the Young Fire Equipment Corporation. This 1932 Ford AA series Case-built pumper is rated at 300-gpm with a 150-gallon water tank. It is one of only five Case-named units built. This unit was delivered 12/19/32.

Pavilion, New York, also owns a rare Case Fire Apparatus manufactured unit. This example shows a 1930 Cadillac touring car conversion rebuilt by Case in 1933 into a rescue squad. It is painted with a striking shade of royal blue and outfitted with full white-wall tires.

East Bloomfield-Holcomb, New York, still has in reserve status this 1934 Diamond-T 211 series pumper. This unit is the earliest recorded Cayasler unit still in existence. It was delivered 5/19/34, only four months after the Cayasler name change came into use. Since its clean lines and professionally styled body counters the early and much more primitive Case units, it may have been subcontracted. It is rated at 350/150.

Frewsburg, New York, has another example of a still surviving early Cayasler unit. This 1934 Dodge was equipped with a Barton U40 pump and dual 150-gallon water tanks. It was one of many such standard-looking units produced in the mid-1930s by this company. This unit was originally painted a cream shade of yellow.

Ridgway, Pennsylvania, purchased this long 235-inch wheel-based city-service ladder unit in 1936. This piece was a unit of many firsts for Cayasler. It was their first attempt at a ladder unit, their first fully metal ground ladder compliment, and the first major use of the Buffalo, New York-built Stewart chassis. This Stewart 49H series was contracted on 3/16/36.

Jamestown, New York, was supplied with this open cab bench seat Cayasler hose wagon in 1937. The unit was also equipped with a PTO booster pump and 100-gallon tank. It was the company's first attempt using the Stewart 58A 3-ton chassis. This clean looking hose wagon is still on the city's roster and is used for PR work.

Fredonia, New York, was the owner of this massive looking Cayasler city-service ladder truck. It was mounted on a 1937 Stewart 58A 3-ton chassis and was the first fully enclosed ladder unit in the country. The unit sported an incredible 250 feet of ground ladders including a 65-foot, three-section extension ground ladder.

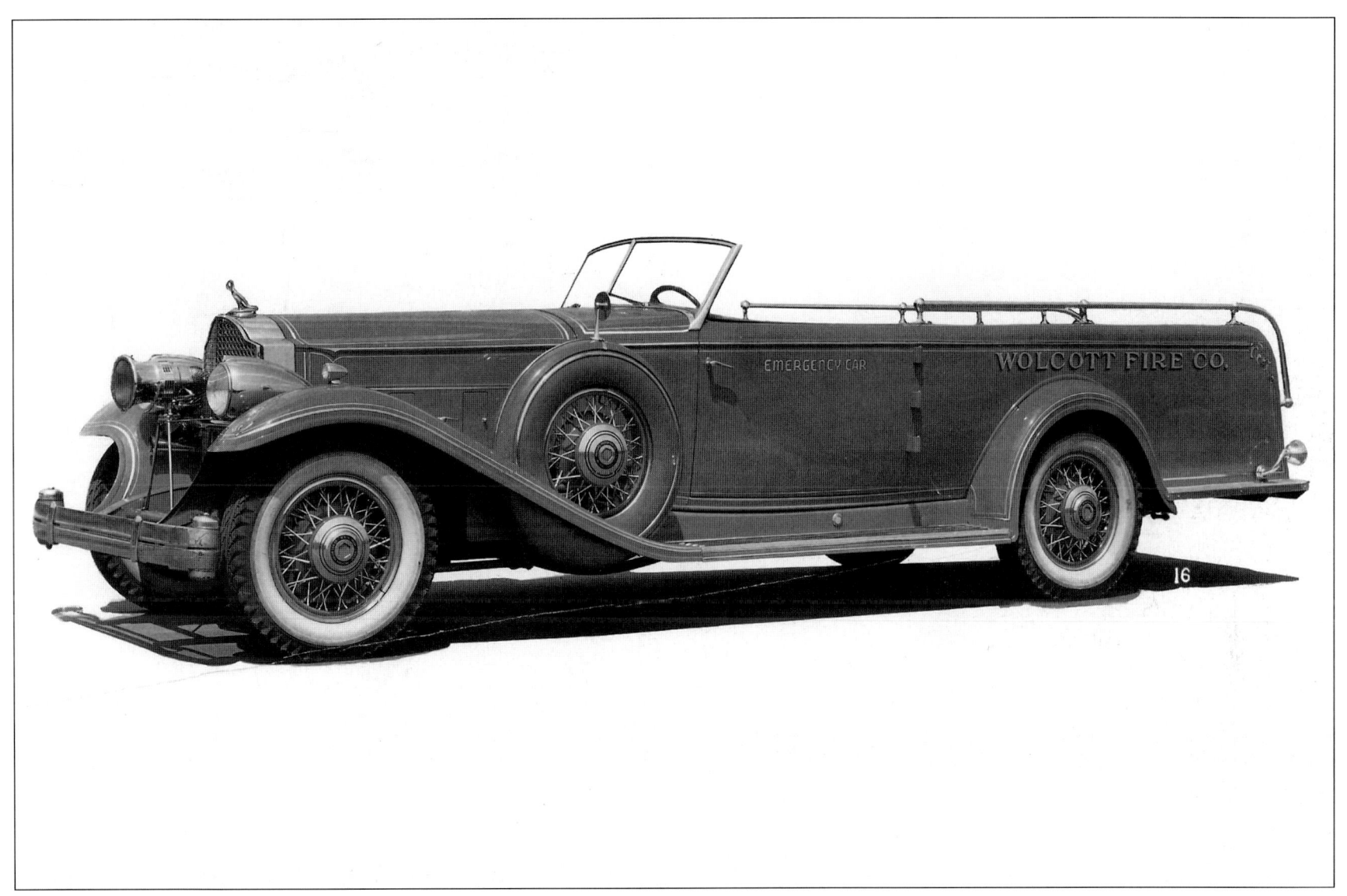

Wolcott, New York, purchased one of the last auto conversions done by this company. Cayasler converted this 1932 Packard touring car into a rescue/booster car in 1937. It was equipped with a 100-gpm Panama pump and 50-gallon booster tank. It also sported full body bench seats.

North Rose, New York, chose this rare and attractive 1938 Reo IDH series chassis for their pumper. The fully enclosed cab unit was rated with a 200-gpm pump and 300-gallon water tank. The oval Cayasler name plate that was used on apparatus from 1934 to 1944 can be seen below the gold leafing on the door.

Wellsville, New York, once again chose Cayasler to provide them with a new piece of apparatus. This chrome nosed 1939 International D30 series city-service ladder unit came equipped with a 200-gpm booster system and a 200-gallon water tank. It was painted a stunning shade of white over Kelly Green, befitting The Emerald Hook & Ladder Company where it served.

Wellsville, New York, was a faithful Cayasler-Young purchaser for many years in the 1930s and 1940s. Shown here is Dyke Street Engine Company 2's chrome cowled 1938 White 710 series 600-gpm pumper. It also had a 200-gallon water tank.

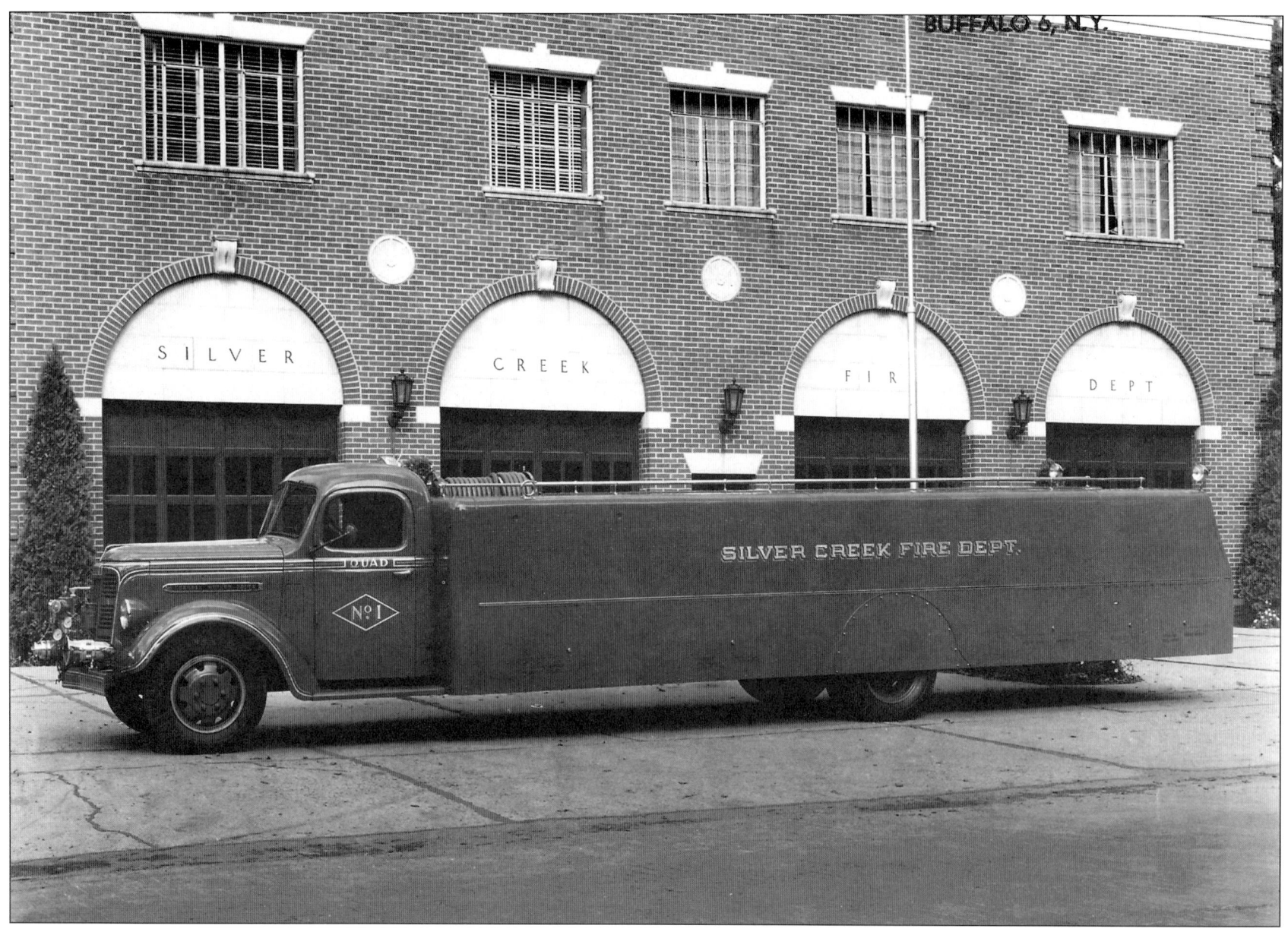

Silver Creek, New York, purchased Cayasler's first fully enclosed-bodied quad. This 1938 GMC model T-23 DSC chassis with extended 232-inch frame was equipped with a 500-gpm front-mounted pump, 150-gallon water tank, and over 200 feet of metal ground ladders.

Eggertsville, New York, chose this rugged looking 1940 Brockway 150X series for their new pumper. This Cayasler was equipped with a 500-gpm pump and 300-gallon tank. The over-the-cab prepiped deck gun and left side-mounted 28-foot extension ladder made it a bit out of the standard norm.

Forestville, New York, placed in service this trim looking 1941 Diamond-T 404 series pumper. It was rated at 500-gpm and had a 200-gallon water tank. The unit served in front line service into the late 1960s.

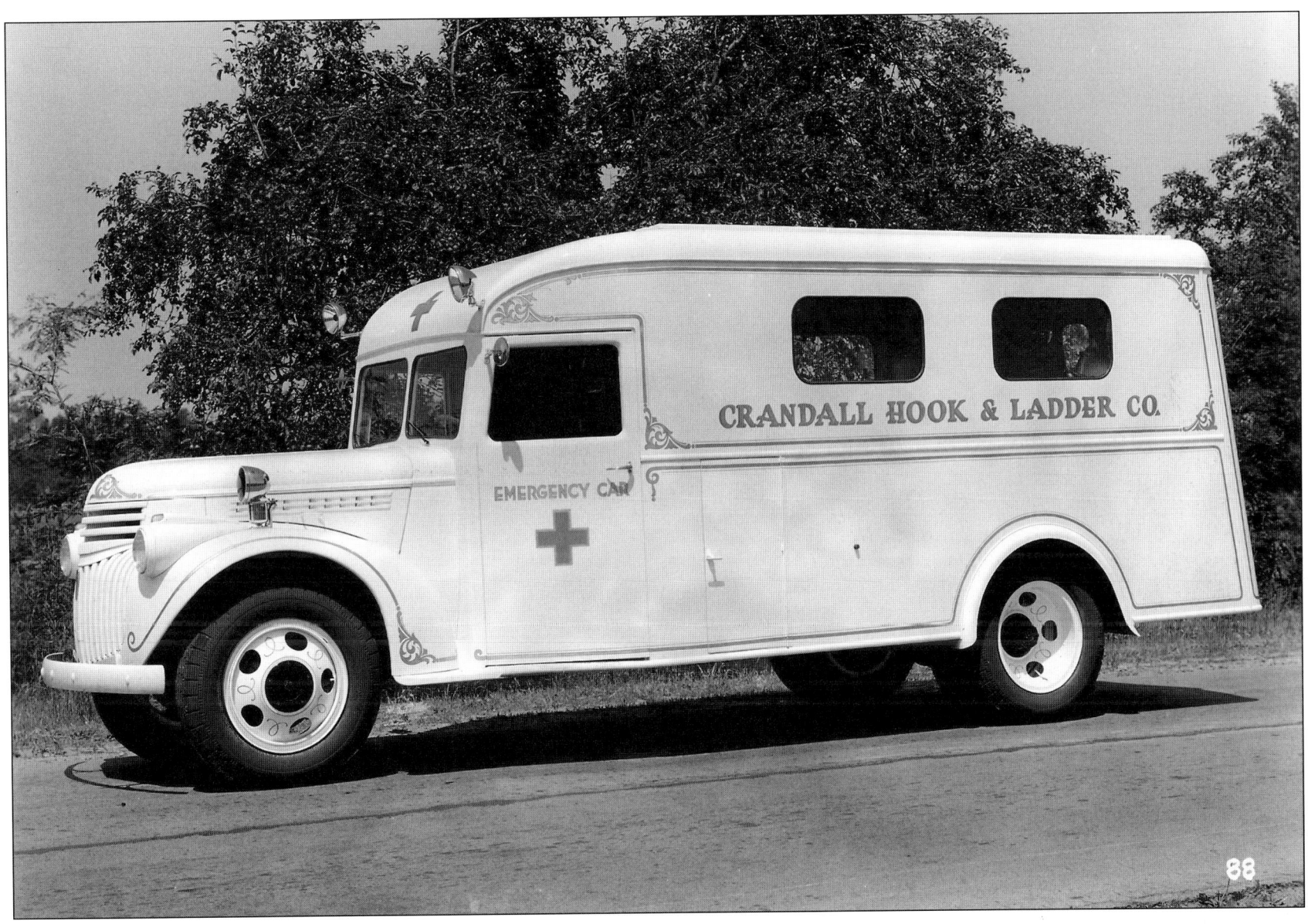

Alfred, New York, placed Cayasler's first attempt at building a heavy rescue in service. This 1942 no chrome war baby Chevrolet YS series chassis was also equipped with a 200-gpm booster system and 150-gallon tank. Its functional rescue body styling and design would serve the entire apparatus industry well into the 1960s. Although assigned to the Crandall Hook & Ladder Company, it carried no ladders.

The U.S. Army during World War II placed many orders for fire apparatus with nearly all companies still producing fire units during the war. Cayasler was no exception. Shown here is a group of five 1942 Ford 2G8T special military chassis 500-gpm pumpers with 300-gallon tanks ready for deployment in full olive drab coloring.

This rare in-plant photo shows eight U.S. Army units being built for the war effort. These 1943 Chevrolet YS series chassis pumpers would provide the Cayasler Company with two much coveted "E" awards for Excellency by the military. These units would also be the last to bear the Cayasler name.

Shown here are the two Army/Navy "E" awards given to Cayasler in 1943 and again in 1944 after the name change to Young. The awards were given for excellence and honor in war production.

Fort Ann, New York, has the distinction of receiving the first piece of apparatus to carry the Young name. This no-chrome war baby 1944 International K series 500-gpm pumper with a 300-gallon water tank was one of very few municipal units built during the war.

Stafford, New York's unit was one of only a few built on a Federal chassis. This 1946 18M series unit was equipped with a 500-gpm pump and 300-gallon tank. Its clean lines set a recognizable body style that would last into the late 1950s.

Chili, New York, purchased this unusual 1946 Ford pumper. It was built with both a 350-gpm front-mount pump and Hardie high pressure fog package at its mid-body. In addition to a 300-gallon water tank, it was equipped with overhead ladder racks. Although it is a 100 percent Young product, it bears a striking resemblance to that of an Oren.

Cheektowaga, New York's Doyle Hose 1 chose a 1947 Mack EH series chassis for their Young bodied 750-gpm pumper. This unit also had a 300-gallon water tank. Young used very few Mack chassis during their sixty-year history.

Alexander, New York, was one of only a handful of departments that purchased Mayday Body and Equipment Corporation fire apparatus. Many of these units were actually built by Young under a subcontract and nearly all were identical in styling to Young. This 1947 International KB series unit had a 350-gpm front-mounted pump and 1500-gallon tanker body.

Victor, New York, purchased this International KB series Young tanker unit in 1948. Note the striking similarity to the Mayday unit mentioned above. Both units also shared identical capacities.

Houghton, New York, purchased this tough looking Brockway 152W series Young pumper in 1947. This 500-gpm with 500-gallon tank unit also sported a Young trademark booster reel system that was placed end to end. This styling was used by Young well into the late 1950s.

Palmyra, New York's 1949 Maxim quad ladder truck was a 100 percent Maxim-built unit. Young sold full Maxim constructed units in the late 1940s in order to gradually ease themselves into an agreement with Maxim to provide custom chassis for the Young Company. It carried Young contract numbers on its ID plate.

Sheridan, New York, ordered this outstanding looking 1950 Diamond-T 420 series. The unit was rated with a 500-gpm pump and 500-gallon water tank. It was painted white with red wheel rims. Its strong, clean, and fine lines were typical of Young styling for this era.

Boston, New York, took delivery of one of Young's first attempts at mini-sized pumpers. This 1950 GMC 150 series pickup chassis served as the base for this compact high-pressure fog unit. It was equipped with a Van Pelt HPF pump and 150-gallon water tank.

East Dunkirk, New York, provided Young with a 1941 WA22 series chassis in 1951. This powerful looking pumper was rated at 500-gpm and had a 600-gallon water tank. And, yes, it is a white, White!

CLYDE FIRE DEPARTMENT

220-A

Clyde, New York, purchased this impressive looking Diamond-T model 920 quad from Young in 1951. It had a 750-gpm mid-mounted pump and a 200-gallon water tank. Young produced many quad and city-service ladder trucks throughout its sixty years. Nearly half of all quads and CSTs had enclosed bodies like this piece.

Westfield, New York, ordered a unit that was the opposite in styling and function to the Clyde, New York, piece previously mentioned. They chose a 1951 Dodge B-3 series chassis with a more traditional open-bodied ladder rack. This city-service ladder unit also sported wooden ground ladders.

Silver Creek, New York, chose a squad body concept for their 1951 GMC HC 644 series pumper. This engine had a 750-gpm pump, 300-gallon water tank, and rear fender skirts in addition to a fully enclosed body. At one time this department had three similar styled Young pumpers in service.

Margaretville, New York, chose this 1952 Dodge B3T chassis for their 500-gpm Young pumper. This heavy looking standard pumper came with a 300-gallon tank, recessed booster reels, and overhead pump panel lights.

Otego, New York, ordered this very functional four-wheel drive Dodge Power Wagon mini-pumper in 1952. Young equipped it with a 350-gpm front-mounted pump, overhead ladder racks, and three pieces of suction hose. This nice looking white-painted compact mini also held 350 gallons of water.

Clifton Springs, New York, purchased this clean lined tanker in 1952. This unit was the only Reo Gold Comet model F20 chassis used in Young production. It was rated at 350-gpm with a 1500-gallon tank. Unlike most Young tankers of this era, this body was flared just behind the pump compartment to give maximum tool storage.

In the early 1950s, Young teamed up with Army's Fire Service of Old Forge, New York, to produce a quick-attack, mini-foam pumper called the "Little Mo." This Young unit was delivered to the New Haven, Connecticut, airport in 1954. In later years, Sanford and American LaFrance would also build units using the "Little Mo" styling components.

The Little Fire Truck with the

BIG PUNCH

"little Mo"

AMERICA'S MOST PRACTICAL FIRE FIGHTER

CITY

MUNICIPAL
AIRPORT
CITY OF NEW HAVEN

NEW HAVEN, CONN.

IDEAL FOR MUNICIPAL · RURAL
INDUSTRIAL · CIVIL DEFENSE and
AIRPORT FIRE PROTECTION

EVERY FIRE DEPARTMENT SHOULD HAVE ONE

GM Detroit Transmission Division, located in Ypsilanti, Michigan, purchased one of Young's "Little Mo" minis in 1953. This unit was mounted on a Chevrolet 4100 series chassis and came equipped with the Army's Fire Service package. This concept proved successful for Young, who would go on to build many such units.

Lancaster, New York, was one of the first departments to order a Maxim custom chassis Young fire body designed unit. Citizen's Hose pumper is a 1953 model 1417 and clearly shows the classic design marriage of Maxim chassis and Young body. It was rated at 750-gpm and had a 300-gallon water tank.

Wendelville, New York, purchased this 1954 Dodge Power Wagon mini with compartments in mind. This unit came equipped with a front cable system, 250-gpm pump, and 250-gallon water tank.

Swormville, New York, sent their 1946 White WA series pumper back to Young in 1955 for rebuilding. These two photos show the modernization done by the factory. The unit was rated at 500-gpm and updated with a 500-gallon tank.

Lake View, New York, was one of many departments who utilized the popular International R196 chassis. The company built many pumpers on this series chassis over a ten-year period. These twin units were delivered in 1955. Both had 750-gpm pumps and 800-gallon tank pieces.

Texaco Laboratories at Beacon, New York, purchased this clean-lined mini in 1955. It was mounted on a Chevrolet 3600 series chassis and was rated at 125-gpm with 125 gallons of water.

Niagara Falls, New York, purchased this demo from Young in 1957. It was built in 1955 as a prototype model for the New York State Thruway System. The Thruway Authority chose a slightly larger scale model for its fleet, and Young won the bid to build twenty-six foam units over a three-year period. This unit is mounted on a Chevrolet 4100 series chassis.

The New York State Thruway chose the larger Chevrolet 6400 series chassis for its fleet order. All twenty-six units were identical with Rockwood foam packages, 125-gpm Van Pelt high-pressure pumps, 330-gallon water tanks, and 25-gallon foam tanks.

The first shipment of New York State Thruway units arrived in 1956. Shown here are five units that were assigned statewide. All of the twenty-six units were painted New York State Blue.

East Fishkill, New York's 1956 Ford F-900 pumper reflects the strong classic lines of the mid-1950s Young era. This unit was rated at 500-gpm with a 1200-gallon water tank. It also came equipped with a Van Pelt high-pressure fog booster system.

GM Corporation Chevrolet Forge Division at Tonawanda, New York, operated this "Youngster" in-plant unit. This 1956 model was rated with a 90-gpm pump, 100-gallon water tank, 200 pounds of CO_2, and 5 gallons of foam. GM in the Buffalo area operated no less than ten such units in its numerous plant system.

Young marketed many of these in-plant units. They were built on a standard Kalamazoo chassis and varied in fire fighting components. Dubbed the "Youngster" model, many were built between 1956 and 1979.

West Seneca, New York, Reserve Fire District 3 purchased this well integrated Maxim/Young in 1957. Young continued to use the Maxim base chassis well into the 1960s. This unit was rated at 750-gpm and had a 600-gallon water tank.

Cheektowaga, New York's U-Crest Fire District 4 also chose a Maxim chassis for their 1957 heavy rescue. This unit was the only Young heavy rescue constructed on the conventionally styled Maxim chassis.

Clarks Summit, Pennsylvania, chose a differently styled Young heavy rescue. This unit shows a non-integrated box without windows, which was mounted on a 1957 Dodge 8 series chassis.

Point Pleasant, New York, purchased Young's first attempt at using the Ford C series cab over. The Ford C series would prove to be the most popular chassis used by Young. Hundreds of units would be produced over a thirty-four year period. This unit is a 1957 model rated at 500-gpm and had a 600-gallon tank.

Ross Township, Pennsylvania's 7th District Fire Company also purchased a unit that would prove very popular among Young customers. The V196 International chassis was used extensively by Young in the late 1950s and 1960s for many units. This piece is a 1958 model rated at 750-gpm with a 600-gallon tank.

Frederick, Maryland's United Fire Company 3 chose this integrated body and chassis for their heavy rescue. Mounted on a Ford F750 chassis, this classic looking unit also sported the Young half cab body. It was delivered in 1958.

Clinton, Maryland, has always been known for their classic styled apparatus. This 1958 GMC model 630 is no exception. It was constructed with many of the Young goodies including chrome cowl, open cab, and back step windshield. The unit was rated with a 750-gpm pump and had a 800-gallon booster tank.

Young did very little advertising during their sixty years. The Young family felt that whatever unit could be shown in an ad would always only reflect on the past. Shown in this rare ad is Ripley, New York's 1958 International V196 pumper. It was rated at 750-gpm with a 600-gallon tank.

YOUNG FIRE EQUIPMENT CO.
1548 WILLIAM ST.
BUFFALO N.Y.

3129

Niagara Falls, New York, purchased one of the last old-styled Maxim chassis Youngs in 1959. This style Maxim base chassis would soon be replaced by the popular "S" series. Young would continue to use Maxim chassis on their custom bids well into the late 1960s. They would stop using Maxim only after the launching of their own custom Crusader series in 1967. Engine 2 was rated at 1000-gpm and had a 300-gallon water tank.

Xerox Corporation of Webster, New York, ordered this compact International B series mini in 1959. The new mini series designed body replaced the round fender series minis used by Young since the 1950s. This piece carried a 250-gpm pump and 200-gallon tank.

![Young Fire Equipment Corporation advertisement showing two Young pumpers — an F series Ford delivered to Savage, Maryland, and a C series, Buffalo 6, N.Y.]

FIRE EQUIPMENT CORPORATION

SAVAGE VOL. FIRE DEPT.

BUFFALO 6, N. Y.

In 1997, The Buffalo Fire Historical Society created this postal commemorative envelope honoring The Young Fire Equipment Corporation. The envelope depicts two Ford hose wagons delivered to Buffalo, New York, in 1964.

This ad, produced in 1960, shows two standard Young pumpers from that era. Shown here is an F series Ford delivered to Savage, Maryland, and a C series provided to Elsmere, New York.

Springdale, Connecticut, serial 60-14. This first snorkel unit originally built in 1960 as a demonstrator on a Ford F model chassis was later sold to Springdale. The 65-foot snorkel served for twenty-five years and in 1985 was transferred to a new Ford C-8000 chassis.

Clinton, Maryland, serial 60-07. Engine 256 was built on a Ford F-600 cab and chassis with 142-inch wheelbase, and was equipped with a Hale CBP 250-gpm pump and a 200-gallon booster tank. Delivered in May 1960, this apparatus now serves in Tylerton, Maryland.

Penfield, New York, serial 61-01. Delivered in 1961, this unique apparatus was built on a Maxim F model cab and chassis. The long wheel base quad was equipped with a Hale 750-gpm pump and a 350-gallon booster tank. The full height rescue style compartments gave this apparatus a powerful fire-fighting punch.

Buffalo, New York, serial 61-25. This apparatus was Young Fire Equipment's second snorkel and was delivered to their home department of Buffalo in 1961. The 85-foot snorkel was mounted on a Ford F-950 chassis and was equipped with 208 feet of ground ladders and a specially designed three-door cab.

LaPlata, Maryland, serial 62-12. Young enjoyed a loyal following in the State of Maryland with many fire departments. This unique pumper, built on an open-cab Mack B model chassis, was equipped with a 750-gpm pump, 500-gallon booster tank, and an extended front bumper for suction hose and warning devices. Young removed the roof and streamlined the body into the side cab panels.

Hillcrest, New York, serial 62-17. Delivered in November 1962 at cost of $48,527, this model 65SFF was the first snorkel quint built by Young. The chassis is an International model V-206 with a 230-inch wheelbase. The apparatus was equipped with a Hale model QLD 750-gpm pump, 175-gallon booster tank, 207 feet of ground ladders, and a hose bed for 1200 feet of 2 1/2-inch hose.

U-Crest Fire Company, Cheektowaga, New York, serial 62-34. This fire company operated a pair of Maxim S model pumpers with Young body work. Engine 4, delivered in 1962, was equipped with a 1000-gpm pump and a 400-gallon booster tank. Note how the pump panel and body were mated with the Maxim cab behind the drivers door. At one time, U-Crest operated six pieces of Young apparatus.

Livingston, New Jersey, serial 63-26. Young did not produce a great number of rescue trucks. This one, delivered in 1963, built on an International model 1700 4x4 chassis, was equipped with a Pitman model PC3D crane rated at 6600 pounds, and used a 7.5 KW power take-off driven generator. This was the first rescue truck produced in the United States with a crane unit.

Clinton, Maryland, serial 64-79. This was the first of seven open-cab Ford C-1000 pumpers built by Young. Constructed on a 175-inch wheelbase, the apparatus cost $37,000 when delivered in August 1964. The apparatus was equipped with a single stage Hale 1000-gpm main pump, a Hale 2CPB high-pressure pump, and a 400-gallon booster tank. A wide array of warning lights and devices adorned the front of this beautiful pumper.

Buffalo, New York, serial 64-82/83. The Buffalo Fire Department operated two Young hose wagons delivered during 1964. Constructed on Ford C-750 chassis with 153-inch wheel bases, these were equally equipped with three Samuel Eastman monitors (one rated at 2000-gpm and two rated at 1100-gpm) and carried 3000 feet of 3 1/2-inch hose. These units remained in service until the mid-1980s.

Shaker Road Fire Company, Loudonville, New York, serial 65-109/81-730R. Constructed on a FWD custom chassis with a 172-inch wheelbase, the apparatus has a two stage Hale 1250-gpm pump and a 300-gallon water tank. Rebuilt by Young during 1982, Engine 448 was re-powered with a Detroit Diesel. A new 500-gallon water tank was installed together with three Mattydale cross lays, Stang gun, and a 5-inch front suction.

Coram, New York, serial 66-137. The Coram apparatus incorporated a 65-foot Pitman snorkel with a 1000-gpm pump and a 350-gallon booster tank. Mounted on a Ford CT-850 chassis, the unit was powered by a 534 cubic-inch gasoline engine through an Allison six-speed automatic transmission. The hose bed could accommodate 1200 feet of 2 1/2-inch hose. Two preconnected 1 1/2-inch attack lines were recessed over the fire pump behind the horizontal rollers. Overall travel height was 9 feet, 10 1/2 inches on this apparatus.

Virginia Beach, Virginia, serial 67-160. The unique Crusader series was introduced in 1967 with this demonstrator pumper that was ultimately delivered to the Princess Anne Courthouse Fire Company. The low profile cab was 93 inches wide and featured dual rear periscope mirrors inside the roof-mounted tunnels. The Madsen Company, in Bath, New York, built the chassis for Young. A total of 43 Crusader series apparatus were produced between 1967 and 1973. This pumper was equipped with a 1000-gpm Hale pump and a 500-gallon booster tank on a 178-inch wheelbase.

Kings Park, New York, serial 67-199. The first order for the new Crusader series came from Kings Park for a pumper, and this 85-foot Pitman Snorkel. Powered by a Waukesha 325 horsepower gasoline engine, this unit was mounted on a 228-inch tandem axle Madsen chassis. The apparatus was equipped with a 1000-gpm Hale fire pump and 200 feet of aluminum ground ladders. The Stang gun in the platform was capable of flowing 1000-gpm with a basket capacity of 900 pounds.

North Greece, New York, serial number 68-222. The Crusader series enjoyed a strong customer following especially in the State of New York where Young delivered many units over the years. The North Greece pumper was powered by a Waukesha F-817 model engine through an Allison HT-70 six-speed automatic transmission. The 1250-gpm Hale QLDA pump was run by a transmission-mounted power take-off that was pioneered by Young. The Budd style wheels were chrome plated to give this pumper a pleasing, balanced appearance.

Allentown Road, Maryland, serial number 68-217. The Allentown Road Fire Department operated five pieces of Young apparatus between 1964 and 1988. Engine 322 was placed into service in December 1968 and is an example of Young's craftsmanship with fire apparatus. The Ford C-850 chassis with a 175-inch wheelbase carried a 1000-gpm pump, a 400-gallon booster tank, and two electric cable reels over the pump. The white-over-black paint scheme was a special treatment that was employed by several fire departments in Maryland.

Kenilworth Fire District of Tonawanda, New York, serial number 68-243. This Crusader pumper was the first of five rescue pumper style bodies constructed by Young. Built on a 176-inch wheelbase Madsen chassis, the 1250-gpm unit featured a 500-gallon booster tank, six enclosed compartments, and a center walkway hose bed design with enclosed ground ladders. Delivered in August 1968, the pumper originally cost $38,977.

Nyack, New York, serial number 68-248. Young produced relatively few rescue squad apparatus; however, those units built had some unique design features. This rescue for the Nyack Fire Patrol, delivered in 1969, was built on a GMC 9700 model, tilt-cab chassis. The rescue body had ten enclosed compartments and was equipped with a 10 KW generator and deck-mounted Mars floodlights. The cab-mounted Solar-Ray warning lights were installed on custom-built chrome brackets on which the Federal siren and bell were mounted.

East Alton, Illinois, serial number 69-289. Two of the forty-three Crusader series units found their way to other body builders. This one was sold to Towers Fire Apparatus of Freeburg, Illinois. The chassis was powered by a Waukesha F-817G gasoline engine with a Hale 1500-gpm fire pump and a 300-gallon booster tank. Towers apparatus featured a unique pump control panel with a deck gun installed at the rear of the engine enclosure.

St. Marys, Pennsylvania, serial number 66-157. The Crystal Fire Department took delivery of this unique 75-foot Pitman snorkel in September 1966. This was the only unit ever built by Young on a Mack C model chassis. The apparatus was equipped with a Hale 2CBP-fire pump, 200-gallon booster tank, and 208 feet of ground ladders. The cab roof had a special notch to provide an overall travel height of 11 feet, 4 inches.

Vienna, Virginia, serial number 70-331. Wagon 2, delivered in 1970, was built on a Duplex model R-CF open-cab chassis. Powered by a Detroit 8V-71N engine, this unit produced 350 horsepower through a Spicer model 6853 five-speed transmission. The 172-inch wheel base chassis carried a Hale 1000-gpm pump and a 500-gallon booster tank. The body featured preconnected hose reels for 1 1/2-inch hose inside the rear compartments and a 600-gpm wagon pipe mounted at the right side of the rear body. This design together with the windshield-mounted Roto-Ray warning lights were common on apparatus in the Washington D.C. area.

Lynbrook, New York, serial number 70-310. This Crusader pumper had many features incorporated into its final design. A Hale model QLDA fire pump was PTO driven and rated at 1000-gpm. A dual foam system with a 40-gallon tank supplied both Mattydale cross lays. The center hose bed walkway led to a rear-facing bench seat over the fire pump that seated three fire fighters. The apparatus wheelbase was 178 inches with an overall length of 28 feet, 1 inch. It was delivered in May 1970; the apparatus contract price was $54,500.

Kodak Park Fire Department, Rochester, New York, serial number 70-333. Young built relatively few Class A apparatus for industrial concerns. The Kodak Park Fire Department operated three stations at their Rochester facility and in 1970 took delivery of this Ford C-950 pumper. Equipped with a 54-foot Squrt, this 1250-gpm pumper carried an arrangement behind the pump panel providing access to a storage area and a preconnected Stang gun rated at 1000-gpm. The chassis was powered by a Ford 534 cubic-inch engine through an Allison MT-42 automatic transmission.

Brookhaven, New York, serial number 70-350. Delivered in December 1970, this tandem axle pumper is the only Crusader unit to be built with a 54-foot Squrt water tower. The apparatus was powered by a Cummins model NTF-365 diesel engine with a Spicer standard transmission on a 228-inch wheelbase. The apparatus body was equipped with a Hale 1000-gpm pump, a 1500-gallon booster tank, and a hose bed for 800 feet of 2 1/2-inch hose. The Squrt boom has a working height of 54 feet, with a horizontal reach of 45 feet while flowing 1000-gpm.

Eden, Pennsylvania, serial number 71-379. This pumper was the first Bison apparatus built and delivered by Young. Due to the high cost of hand building the Crusader cab, Young conceived the idea of producing a lower cost unit using a Cincinnati cab. Originally produced as a demonstrator pumper, it was later sold to the Eden Fire Company near Lancaster, Pennsylvania. The 178-inch wheel base chassis was built by Madsen using a low profile cab, Ford 534 cubic-inch gasoline engine, and an Allison MT-42 automatic transmission. The Hale QLD pump was rated at 1000-gpm, and the booster tank capacity was 500 gallons.

Allentown Road, Maryland, serial number 72-391. This open-cab 1000-gpm Bison pumper is probably the most unique apparatus ever produced by Young. The roof was removed from a low profile Cincinnati cab and special reinforcements were installed across the front of the cab down from the center of the windshield. A 350 horsepower 8V-71N Detroit diesel powered the Madsen-built chassis. Special options included a special deep FDNY style hose bed, cab door-mounted horns, 5-inch front suction, polished aluminum wheels, and a windshield-mounted Mars light. This apparatus cost $59,800 when delivered to Allentown Road in September 1971.

Eastman Kodak Company, Windsor, Colorado, serial number 71-402. This pumper, delivered in 1971, went to the Eastman Kodak's Colorado facility. The Ford C-900 chassis carried a 1250-gpm pump, 500-gallon booster tank, and a 50-foot Telesqurt on a 175-inch wheelbase. A booster reel and a Stang gun were carried over the pump and were accessed from a special step arrangement behind the pump panel.

Erlton Fire Company, Cherry Hill, New Jersey, serial number 71-394. The State of New Jersey was a fertile territory for Young fire apparatus. This Bison 1500-gpm pumper was heavily decorated with gold leaf lettering and striping. The body sides were higher than normal to accommodate the 1000-gallon booster tank and a large hose bed. The apparatus was powered by a Detroit Diesel 8V-71N engine through an Allison HT-740 transmission.

New Hackensack, New York, serial number 74-513. Young used the ubiquitous C series Ford chassis as the platform for many pieces of apparatus. In 1974 Young designed the Custom Power series of apparatus using Detroit Diesel engines. New Hackensack received this 1000-gpm pumper, which was powered by a 350 horsepower 8V-71N engine through an Allison HT-740 four-speed automatic transmission.

Pearl River, New York, serial number 72-458. The Pearl River Hook and Ladder Company acquired this Bison 1250-gpm pumper in December 1972. The 200-inch wheelbase Madsen chassis was powered by a Waukesha model F817G gasoline engine. The apparatus body featured a walkway in the rear hose bed. Chrome plated front fenders and extensive gold leaf striping gave this apparatus a well-balanced appearance.

Nanuet, New York, serial number 73-468. The Nanuet Fire Department was a loyal customer for Young. Walter Motor Truck of Voorheesville, New York was now producing the Bison chassis for Young. This unit was powered by a Cummins model NTF-365 diesel engine through an Allison HT-70 six-speed automatic transmission. Due to the length of this in-line engine, the chassis was built on a special 187-inch wheelbase. The apparatus was equipped with two Hale pumps; a QSMD 1500-gpm, and a 2CBP high-pressure pump that supplied a rear-mounted booster reel and a gated discharge inside the front left compartment.

Willow Street, Pennsylvania, serial number 73-481. Young built two 4x4 Bison pumpers with Walter chassis. This one, for Willow Street, was powered by a Cummins V-903 320 horsepower diesel engine through an Allison HT-70 automatic transmission. The apparatus was equipped with a Hale two-stage 1500-gpm pump and a 750-gallon booster tank. The apparatus was delivered with 18.0 x 19.5 single flotation tires on the rear axle and later changed to 11.0 x 20 dual tires.

Chews Landing, New Jersey, serial number 73-492. This 85-foot snorkel is probably one of the most ornate snorkel platforms ever placed in service. The Walter chassis, with a 228-inch wheelbase, was powered by a Detroit diesel 8V-71N engine providing 350 horsepower. The apparatus is equipped with a 1500-gpm pump, 300-gallon booster tank, and 208 feet of ground ladders.

White Oak, Pennsylvania, serial number 73-488. Young produced six 50-foot Telesqurt apparatus on their Bison chassis. This one, for the Rainbow Fire Company, was delivered in September 1973 at a cost of $85,493. The Walter Motor Truck chassis was powered by a Cummins NTF-365 diesel engine with an Allison HT-740 automatic transmission. The apparatus was equipped with a Hale 1000-gpm pump, a 500-gallon booster tank, three Mattydale cross lays, and a hose bed for 1600 feet of 3-inch hose.

Tallman, New York, Serial Number 73-494. The last Crusader apparatus was delivered to Tallman, New York, in December 1973. The 65-foot Pitman snorkel was installed on a Walter chassis with a 228-inch wheelbase. The apparatus had an overall height of 10 feet and an overall length of 40 feet, 10 inches. A Hale 1250-gpm pump supplied the snorkel through a 4-inch waterway. The ground ladder compliment totaled 196 feet with a hose bed capacity for 400 feet of 2 1/2-inch hose and 600 feet of 1 1/2-inch hose.

Morristown, New Jersey, serial number 74-522. Between 1960 and 1974, Young built fifty snorkels in various configurations using many chassis manufacturers. The Resolute Hook and Ladder Company took delivery of the last Young snorkel produced with this 85-foot unit. The Bison cab was mounted on a Walter chassis with a wheelbase of 228 inches. The apparatus carried 194 feet of ground ladders and had a 12.0 KW Onan diesel generator on board.

Erlton Fire Company, Cherry Hill, New Jersey, serial number 76-553. Innovative ideas were engineered throughout this new King Cobra snorkel design. Conceived by Dick Young in 1975 as a result of years of aerial platform building, the King series of aerial devices utilized a Hendrickson FTCOF-2070 model mid-engine chassis with a crankshaft-driven Hale 1250-gpm fire pump. The platform device, built in England, was a Simon snorkel rated at 103 feet working height.

South Line Fire District, Cheektowaga, New York, serial number 76-554. Young formed a working agreement with Ladder Towers Incorporated and began to produce a series of various aerial and platform devices. This apparatus, delivered in 1976 to South Line, featured an LTI 85-foot tower on a Hendrickson chassis with a 228-inch wheelbase. It was powered by a midship Detroit Diesel 8V-71N rated at 350 horsepower through an Allison HT-740 automatic transmission. The body compartments were a modular bolt-on design.

St. Paul Boulevard Fire Department, Rochester, New York, serial number 76-568. The fire fighting power of the larger aerial devices were the springboards for the Fire Commander series. This quint incorporated an LTI 65-foot steel, three-section aerial on a Ford CT-8000 chassis with a 200-inch wheelbase. The Hale 1000-gpm pump had top-mount controls where the operator could stand to control all aerial ladder and pump functions.

Hillcrest, New York, serial number 76-613. Young produced 37 complete Bison apparatus between 1971 and 1976. The last Bison was delivered to the Hillcrest Fire Company in Rockland County, New York. The Walter chassis had a special 209-inch wheelbase and was powered by a Detroit diesel 8V-71N engine. The long body carried 217 feet of ground ladders, Hale 1250-gpm pump, 500-gallon booster tank, and an Onan 6.0 KW diesel generator. The cab was built with Crusader style, roof-mounted mirror tunnels with strobe lights mounted on the front of the cab.

District of Columbia, Engine 13's wagon. The Washington D.C. Fire Department operated nine Ford C-900 series Ward LaFrance pumpers that suffered premature body deterioration due to rust. These units ran as the wagon of a two-piece engine company. During 1979 and 1980, Young rebuilt all nine pumpers using the existing cab and chassis as well as the crew cab and the 750-gpm fire pump. Young fabricated a modular galvaneal steel body, 500-gallon booster tank, and new pump panel with two cross lays. This body design would later be called the Valumaster series. Engine 13's wagon carried DCFD shop number 121. Young assigned no serial number to these rebuilt DCFD rigs.

LaPlata, Maryland, serial number 79-726. Young built the body on this Mack CF midship aerial ladder under subcontract for Mack Fire Apparatus. The 100-foot midship American LaFrance ladder was a one-off project for Young, who had previous experience building bodies on snorkel devices. This aerial chassis had a wheelbase of 228 inches and carried Mack serial number CF-686FCA-1364. Truck 1 was equipped with eight equipment compartments and 165 feet of ground ladders. Warning lights included twin Mars lights and four cab-mounted strobe lights.

Arlington County, Virginia, serial number 81-732. Young rebuilt eight Ward LaFrance custom pumpers for Arlington County between 1980 and 1985. Young delivered three new Duplex D-300 pumpers during 1981, including this one to Engine 77. These units were powered by Detroit Diesel 6V-92TA engines through Allison HT-740 four-speed automatic transmissions. The Value Master galvaneal steel bodies carried a preconnected 1 3/4-inch attack line in the front bumper and two Mattydale cross lays.

Town Line Fire Company, Lancaster, New York, serial number 82-745. The first Crusader II pumper was delivered to the Town Line Fire Department, who were first due at the Young plant on Cemetery Road. This innovative mid-engine pumper was based upon a Duplex D-350 chassis with a 175-inch wheelbase. The 1250-gpm fire pump was mounted under the front of the cab and was crankshaft-driven by the Detroit Diesel 6-71T 275 horsepower engine. All fire pump valves were hydraulic controlled with 4-inch intakes and discharge on all four sides of the pumper. The galvaneal steel body was a modular bolt-on design with eight enclosed compartments.

106

Baileys Cross Roads Fire Company, Fairfax, Virginia, serial number 84-765. This unique pumper was the first enclosed four-door, raised roof pumper delivered in the United States. Wagon 10 is powered by a Detroit Diesel 6V-92TA 350 horsepower engine through an Allison HT-740 transmission. The apparatus was equipped with a 1250-gpm pump, 500-gallon booster tank, and a 40-gallon foam tank. This unit was set with eight preconnected attack lines and carried 1500 feet of 3-inch supply line.

Ellwood Fire Company, Town of Tonawanda, New York, serial number 66-127/84-778. This 75-foot snorkel was originally built in 1968 on an open-cab Ford CT-850 chassis. In 1984, Young rebuilt the snorkel using the original body remounted on a Duplex D-450 cab and chassis. A new Hale 1250-gpm pump was installed with hydraulic valve controls. The apparatus is powered by a Detroit Diesel 6V-92TA engine through an Allison HT-740 automatic transmission.

Haiti, serial number 84-789. Young produced several units for export, including four tankers for Haiti. Built on Chevrolet C-70 chassis, these units were equipped with 250-gpm power take-off driven pumps and 1200-gallon booster tanks. A similar unit was delivered to Gansevoort, New York, using the same specifications. Many of the export apparatus did not carry Young nameplates, as local companies would detail the apparatus and supply hose and equipment prior to delivery.

Bethpage, New York, serial number 84-797/798. These Crusader IIs were a pair of twins that were delivered to Bethpage, New York. The mid-engine Duplex chassis were built on a 175-inch wheelbase with a Young-built four-door cab enclosure. The apparatus were equipped with a 1250-gpm pump, 500-gallon booster tank, preconnected Akron deck gun, and a rear-mounted booster reel. They were both later sold to Newton-Abbott Fire Company of Hamburg, New York, who refitted them to meet their needs.

Spring Valley, New York, serial number 86-810. Young produced several Crusader II pumpers with 1000-gallon booster tanks. This one, for the Columbian Engine Company, was built on a Duplex D-350 two-door chassis with a 191-inch wheelbase and was powered by a 6V92TA Detroit Diesel engine. This apparatus was outfitted with an 8.0 KW hydraulic generator, body and cab strobe lights, and a preconnected deck gun behind the top-mounted pump controls.

Coram, New York, serial number 85-818. This unique apparatus was designed by Young to produce large quantities of foam solution. The unit was equipped with a foam proportioning system, 800-gallon foam concentrate tank, and a 500-gpm foam turret. The chassis was a Ford F-7000 four-wheel drive with a Marmon Harrington front driving axle. The aluminum tread plate body carried two beds for 2 1/2-inch hose and 1 1/2-inch hose for foam delivery. Over the years, Coram operated six pieces of Young apparatus.

Bowmansville, New York, serial number 86-849. Young delivered twin 1250-gpm pumpers to Bowmansville during 1986. These featured the Olympic Gold series of molded fiberglass bodies with six equipment compartments. The chassis was the popular Ford C-8000 with a 175-inch wheelbase. The apparatus was equipped with a 1000-gallon booster tank, three-speed lay attack lines, and a hose bed walkway in the rear. Young produced over eighty-five units with the Olympic Gold body design.

Arlington County, Virginia, serial number 86-860. After rebuilding several pumpers for Arlington County, Young produced this unique tractor-drawn aerial ladder. The trailer is a 1965 American LaFrance 100-foot unit that had new compartments installed together with extensive remodeling of the trailer sub-frame and tiller assembly. A four-door International model S-1954 chassis was modified by Young to accept the fifth wheel, and new bodywork was fabricated to accommodate a Hurst tool and ladder pipe supply line.

Lancaster, New York, serial number 87-870. The village of Lancaster acquired three Ward LaFrance pumpers in 1973. Young rebuilt each of these 1250-gpm units, completing Engine 2 in 1987. New Olympic Gold fiberglass bodies were installed together with 600-gallon booster tanks. The pump configuration was changed to a top-mount design with two-speed lay attack lines. Young rebuilt the original cab and installed rectangular quad headlights and warning lights. The entire project cost $84,517 when completed.

Nanuet, New York, serial number 87-875. This Crusader II 1250-gpm pumper for Nanuet featured a rescue pumper style body with a 500-gallon booster tank. The galvaneal steel body was mounted on a Duplex D-350 chassis with a Young-built four-door raised roof cab. The apparatus is powered by a Detroit Diesel 6V-92TA engine through an Allison HT-740 automatic transmission. The apparatus is equipped with a 10 KW hydraulic generator and carries five preconnected attack lines. When delivered in October 1987, the apparatus cost $223,862.

Chincoteague, Virginia, serial number 88-884. Originally built in 1968 carrying Young serial number 68-220, Chincoteaque's pumper was the sixth Crusader produced. Young extensively rebuilt this unit with an Olympic Gold body, 1000-gallon booster tank, and top-mount hydraulic valve pump control. The pumper is equipped with quad warning lights on the cab front, polished aluminum wheels, top-mounted deck gun, and rear-tank jet dump.

Palm Beach County, Florida, serial number 88N886. This Crusader II pumper is part of fourteen identical units that were delivered to Palm Beach County. These Duplex chassis pumpers were built on a 186-inch wheelbase and were powered by Cummins L-10 300 horsepower engines. The apparatus was equipped with 1250-gpm pumps, 750-gallon booster tanks, and three Mattydale cross lays. Engine 33 was one of the first of the groups delivered. This was one of the largest orders for apparatus produced by the Young Corporation.

Miami, Florida, serial number 90N942. Hose 6 is one of six Crusader II pumpers delivered to Miami during 1990. Equipped with a 1250-gpm pump and 500-gallon booster tank, this apparatus featured bus-style crew doors, fiberglass modular body, and a hose bed walkway.

Miami, Florida, serial number 90N925. Miami also took delivery of three Crusader II quints with LTI 75-foot Aqua-Stick ladders. Built on 198-inch wheelbase Duplex chassis, these units were equipped with 1250-gpm pumps and 300-gallon booster tanks. These units were the last group of aerial ladder devices that were produced by Young.

Wayne Township, Indiana, serial number 90N970. This pumper is one of six Crusader II pumpers delivered to Wayne Township during 1990. Engine 21 is the last custom Young apparatus produced. This apparatus was equipped with a 1500-gpm pump, 1000-gallon booster tank, 60-gallon foam tank, and cab-mounted hydraulic valve pump controls. Originally painted in pearl white and forest green colors, this apparatus was repainted in 1998 and is now white over red.

Villas, New Jersey, serial number 91-987. This pumper is the last complete Young apparatus produced. The Ford C model chassis is from serial number 72-382, and a new fiberglass body and water tank were installed. Delivered on March 27, 1991, this apparatus was an excellent example of the workmanship that was the hallmark of Young.

COMPLETE CRUSADER LIST 1967-1973

Location	Number	Spec
Princess Anne Courthouse, Virginia Beach	67-160	1000/500
Kings Park, New York	67-198	120/500
Kings Park, New York	67-199	85' snorkel/1250/0
Elmont, New York	68-206	85' snorkel
Port Washington, New York	68-211	1000/500
Chincoteague, Virginia	68-220	1000/750
Henrietta, New York	68-221	85' snorkel/1000/200
North Greece, New York	68-222	1250/500
Dansville, New York	68-238	1000/1000
Kenilworth FC, Tonawanda, New York	68-243	1250/500
St. Paul Boulevard FC, Rochester, New York	68-224	85' snorkel/1250/500
Doyle FC, Cheektowaga, New York	68-258	1250/500
Silver Lake, New York	68-264	1000/1000
Orangeburg, New York	69-272	1000/500
Elsmere, New York	70-275	1000/500
Elsmere, New York	70-276	85' snorkel
West Islip, New York	68-278	1250/500
West Islip, New York	68-279	1250/500
Harrison, New York	69-281	1000/500
Nanuet, New York	69-285	75' snorkel
Kempsville FC, Virginia Beach	69-286	1000/500
East Alton, II chassis only	69-289	Towers Body 1500/300
West Babylon, New York	70-291	75' snorkel
Centereach, New York	70-293	85' snorkel/1250/0
Ardsley, New York	70-303	1250/500
Montville, New Jersey	70-304	1000/500
Morris Twp., New Jersey	70-305	1000/500
Lynbrook, New York	70-310	1250/500
Marlton, New Jersey	70-317	1500/500
New Hyde Park, New York	70-336	1250/500
Rescuse House FC, Cheektowaga, New York	70-338	1250/500
Brookhaven, New York	70-350	1000/1500, Squrt
West Ridge, Pennsylvania	70-354	75' snorkel
East Meadow, New York	70-355	85' snorkel
Westhampton Beach, New York	71-366	1000/500
Lake Mohegan, New York	71-369	1250/500
Lake Mohegan, New York	71-370	1250/500
Lakewood, Ohio (chassis only)	71-376	85' snorkel
Millbrook FC, Randolph Twp., New Jersey	71-378	1000/500
Penfield, New York	72-418	1250/500
Hempstead, New York	72-445	85' snorkel
Tallman, New York	73-494	75' snorkel/1250/300
Marlton, New York	73-498	85' snorkel

COMPLETE CRUSADER II LIST 1982-1990

Location	Number	Spec
Town Line, New York	82N745	1250/750
Seaside Heights, New York	82N758	1250/750
Baileys X-Roads, Virginia	84N765	1250/500
Arlington County, Virginia	84N794	1250/500
Bethpage, New York	84N797	1250/500
Bethpage, New York	84N798	1250/500
Spring Valley, New York	86N810	1250/1000
Rainbow Lakes FC, Denville, New Jersey	85N821	1250/500
South Line FC, Cheektowaga, New York	86N831	1250/750
Bellmore, New York	86N840	1250/750
Arlington County, Virginia	88N845	1250/500
Nanuet, New York	88N873	1250/500
Madeira Beach, Florida	88N883	1250/500
Palm Beach County, Florida	88N886	1250/750
Palm Beach County, Florida	88N887	1250/750
Palm Beach County, Florida	88N888	1250/750
Palm Beach County, Florida	88N889	1250/750
Palm Beach County, Florida	88N890	1250/750
Coral Gables, Florida	88N896	1250/750
Coral Gables, Florida	88N897	1250/750
Palm Beach County, Florida	88N900	1250/750
Palm Beach County, Florida	88N901	1250/750
Palm Beach County, Florida	88N902	1250/750
Palm Beach County, Florida	88N903	1250/750
Palm Beach County, Florida	88N904	1250/750
Palm Beach County, Florida	88N905	1250/750
Palm Beach County, Florida	88N906	1250/750
Palm Beach County, Florida	88N907	1250/750
Palm Beach County, Florida	88N908	1250/750
Arlington County, Virginia	88N909	1250/500
Arlington County, Virginia	88N910	1250/500
Coral Gables, Florida	88N911	1250/500
Arlington County, Virginia	89N918	1250/500
Miami, Florida	90N922	1250/500
Miami, Florida	90N923	1250/500
Miami, Florida	90N924	1250/500
Miami, Florida	90N925	1250/300 75' RM -LTI
Miami Beach, Florida	90N930	1250/500
Miami Beach, Florida	90N931	1250/500
Miami Beach, Florida	90N932	1250/500
Wrights Corners, New York	90N936	1250/500
Miami, Florida	90N940	1250/500
Miami, Florida	90N941	1250/500
Miami, Florida	90N942	1500/500
Miami, Florida	90N943	1250/300 75' RM -LTI
Miami, Florida	90N944	1250/300 75' RM -LTI
Wayne, Twp., Indiana	90N945	1500/1000
Wayne, Twp., Indiana	90N946	1500/1000
Wayne, Twp., Indiana	90N947	1500/1000
Arlington County, Virginia	90N951	1250/500
Coram, New York	90N955	1250/500
Arlington County, Virginia	90N965	1250/500
Wayne, Twp., Indiana	90N968	1500/1000
Wayne, Twp., Indiana	90N969	1500/1000
Wayne, Twp., Indiana	90N970	1500/1000

COMPLETE BISON LIST 1971-1976

Location	Number	Spec
Eden, Pennsylvania	71-379	1000/500
Allentown Road, Maryland	71-390	1000/300
Englewood, Colorado	72-01	50'sn/1250/500
Cahokia, Illinois	72-02	1250/500
Secacus, New Jersey	72-03	85' snorkel/100/200
Rockaway Neck FC, Parisipanny, New York	72-391	1250/500
Oceanside, New York	72-392	1250/500
Succasuna FC, Roxbury Twp., New Jersey	72-393	50'sn/100/500
Erlton FC, Cherry Hill, New Jersey	72-394	1500/1000
Elsmere, New York	71-395	1250/750
Whaton, New Jersey	72-396	1000/500
Flanders FC, Cherry Hill, New Jersey	72-413	1000/750
East Aurora, New York	72-420	1250/750
West Lake, Pennsylvania	72-425	50'sn/1250/500
Gaithersburg, Maryland	72-426	1250/500
Pine Hill FC, Cheektowaga, New York	72-427	50'sn/1250/500
Deer Park FC, Cherry Hill, New Jersey	72-443	50'sn/1500/500
Gasport, New York	72-447	50'sn/1250/500
Parsippany FC, Lake Hiawatha, New Jersey	72-453	85' snorkel
Pueblo, Colorado	72-457	1250/500
Pearl River, New York	72-458	1000/300
Batavia Twp., New York	72-463	1500/300
Nanuet, New York	73-468	85' snorkel/1000/200
Verplank, New York	73-474	1000/1000
Longwood, Pennsylvania	73-476	1500/1000
Willow Street, Pennsylvania	73-481	50'sn/1250/500
White Oak, Pennsylvania	73-488	85' snorkel/1500/300
Chews Landings, New Jersey	73-492	1250/500
Main Transit FC, Amherst, New York	73-495	1250/500
Lincroft, New Jersey	73-501	1250/500
Packanack Lake FC Wayne, New Jersey	73-504	1250/500
Hilltop FC, Netcong, New Jersey	74-518	85' snorkel
Morristown, New Jersey	74-522	1250/500
Perry FC, Cinnaminson, New Jersey	74-541	1250/500
Butler, New Jersey	74-543	1250/500
Butler, New Jersey	74-544	100'rm
Pennsauken, New Jersey	74-551	1250/1000
Berlin, New Jersey	76-561	1250/500
Montville, New Jersey	76-610	1500/500
Haddon FC, Westmount, New Jersey	76-611	1250/500
Hillcrest, New York	76-613	

This plant at 204 Cemetery Road in Lancaster, New York, served as the main manufacturing facility from 1966 through 1991. The office area is in the foreground with the main manufacturing and service area in the rear. Behind the plant was a pump test facility where all pumpers underwent the standard tests.

CONCLUSION

Richard Young could best be described as an innovator. The consummate, visionary Dick lead Young Fire Equipment Corporation through several decades of growth and recognition as one of the fire apparatus industry's premier manufacturers. During the 1960s, Young pioneered the concept of snorkel elevating platforms in the northeast and produced the company's first custom chassis apparatus with the Crusader series.

Hundreds of pieces of apparatus were produced on the ubiquitous Ford C model commercial chassis. Young produced the first of these models with such innovations as a full width canopy cab, modular compartment body, and Detroit Diesel power.

Entering the decade of the 1970s, Young continued to introduce innovative and unique apparatus designs. The Bison series of custom apparatus were produced in many configurations including pumpers, Telesqurts and snorkels. Young introduced the first mid-engine apparatus in 1976, and these units help propel Young into the national spotlight. The engineering design concepts embodied in these units were years ahead of what the industry was then contemplating.

A devastating strike in 1977 put the future of the company in jeopardy, and only through the hard work of the Young family, the company survived. The era of the 1980s produced many unique apparatus rebuilding projects for new and past customers. Young introduced the mid-engine Crusader II series apparatus in 1982, and this design together with a modular fiberglass body design formed the backbone of the company's offerings during this period. Young produced forty-five Crusader II apparatus between 1982 and 1990.

Problems with adamant union leadership caused the unfortunate demise of the company. The final apparatus delivered to Villas, New Jersey, left the Cemetery Road plant on March 27, 1991.

"Exceptional quality, value, and performance is engineered into every body built by Young." This quotation from a piece of Young literature describes the valued position the Young Fire Equipment Corporation had on history.

Tom W. Shand

Index

More Titles from Iconografix:

AMERICAN CULTURE

AMERICAN SERVICE STATIONS 1935-1943 PHOTO ARCHIVE ISBN 1-882256-27-1
COCA-COLA: A HISTORY IN PHOTOGRAPHS 1930-1969 ISBN 1-882256-46-8
COCA-COLA: ITS VEHICLES IN PHOTOGRAPHS 1930-1969 ISBN 1-882256-47-6
PHILLIPS 66 1945-1954 PHOTO ARCHIVE ISBN 1-882256-42-5

AUTOMOTIVE

CADILLAC 1948-1964 PHOTO ALBUM ISBN 1-882256-83-2
CAMARO 1967-2000 PHOTO ARCHIVE ISBN 1-58388-032-1
CORVETTE THE EXOTIC EXPERIMENTAL CARS, LUDVIGSEN LIBRARY SERIES ISBN 1-58388-017-8
CORVETTE PROTOTYPES & SHOW CARS PHOTO ALBUM ISBN 1-882256-77-8
EARLY FORD V-8S 1932-1942 PHOTO ALBUM ISBN 1-882256-97-2
IMPERIAL 1955-1963 PHOTO ARCHIVE ISBN 1-882256-22-0
IMPERIAL 1964-1968 PHOTO ARCHIVE ISBN 1-882256-23-9
LINCOLN MOTOR CARS 1920-1942 PHOTO ARCHIVE ISBN 1-882256-57-3
LINCOLN MOTOR CARS 1946-1960 PHOTO ARCHIVE ISBN 1-882256-58-1
PACKARD MOTOR CARS 1935-1942 PHOTO ARCHIVE ISBN 1-882256-44-1
PACKARD MOTOR CARS 1946-1958 PHOTO ARCHIVE ISBN 1-882256-45-X
PONTIAC DREAM CARS, SHOW CARS & PROTOTYPES 1928-1998 PHOTO ALBUM ISBN 1-882256-93-X
PONTIAC FIREBIRD TRANS-AM 1969-1999 PHOTO ALBUM ISBN 1-882256-95-6
PONTIAC FIREBIRD 1967-2000 PHOTO HISTORY ISBN 1-58388-028-3
STUDEBAKER 1933-1942 PHOTO ARCHIVE ISBN 1-882256-24-7
STUDEBAKER 1946-1958 PHOTO ARCHIVE ISBN 1-882256-25-5

BUSES

THE GENERAL MOTORS NEW LOOK BUS PHOTO ARCHIVE ISBN 1-58388-007-0
GREYHOUND BUSES 1914-2000 PHOTO ARCHIVE ISBN 1-58388-027-5
*MACK® BUSES 1900-1960 PHOTO ARCHIVE** ISBN 1-58388-020-8
TRAILWAYS BUSES 1936-2001 PHOTO ARCHIVE ISBN 1-58388-029-1

EMERGENCY VEHICLES

AMERICAN LAFRANCE 700 SERIES 1945-1952 PHOTO ARCHIVE ISBN 1-882256-90-5
AMERICAN LAFRANCE 700 SERIES 1945-1952 PHOTO ARCHIVE VOLUME 2 ISBN 1-58388-025-9
AMERICAN LAFRANCE 700 & 800 SERIES 1953-1958 PHOTO ARCHIVE ISBN 1-882256-91-3
AMERICAN LAFRANCE 900 SERIES 1958-1964 PHOTO ARCHIVE ISBN 1-58388-002-X
CLASSIC AMERICAN AMBULANCES 1900-1979 PHOTO ARCHIVE ISBN 1-882256-94-8
CLASSIC AMERICAN FUNERAL VEHICLES 1900-1980 PHOTO ARCHIVE ISBN 1-58388-016-X
CLASSIC SEAGRAVE 1935-1951 PHOTO ARCHIVE ISBN 1-58388-034-8
FIRE CHIEF CARS 1900-1997 PHOTO ALBUM ISBN 1-882256-87-5
LOS ANGELES CITY FIRE APPARATUS 1953 - 1999 PHOTO ARCHIVE ISBN 1-58388-012-7
*MACK MODEL B FIRE TRUCKS 1954-1966 PHOTO ARCHIVE** ISBN 1-882256-62-X
*MACK MODEL C FIRE TRUCKS 1957-1967 PHOTO ARCHIVE** ISBN 1-58388-014-3
*MACK MODEL CF FIRE TRUCKS 1967-1981 PHOTO ARCHIVE** ISBN 1-882256-63-8
*MACK MODEL L FIRE TRUCKS 1940-1954 PHOTO ARCHIVE** ISBN 1-882256-86-7
NAVY & MARINE CORPS FIRE APPARATUS 1836 -2000 PHOTO GALLERY ISBN 1-58388-031-3
PIERCE ARROW FIRE APPARATUS 1979-1998 PHOTO ARCHIVE ISBN 1-58388-023-2
SEAGRAVE 70TH ANNIVERSARY SERIES PHOTO ARCHIVE ISBN 1-58388-001-1
VOLUNTEER & RURAL FIRE APPARATUS PHOTO GALLERY ISBN 1-58388-005-4
WARD LAFRANCE FIRE TRUCKS 1918-1978 PHOTO ARCHIVE ISBN 1-58388-013-5
YOUNG FIRE EQUIPMENT 1932-1991 PHOTO ARCHIVE ISBN 1-58388-015-1

RACING

GT40 PHOTO ARCHIVE ISBN 1-882256-64-6
INDY CARS OF THE 1950s, LUDVIGSEN LIBRARY SERIES ISBN 1-58388-018-6
INDIANAPOLIS RACING CARS OF FRANK KURTIS 1941-1963 PHOTO ARCHIVE ISBN 1-58388-026-7
JUAN MANUEL FANGIO WORLD CHAMPION DRIVER SERIES PHOTO ALBUM ISBN 1-58388-008-9
LE MANS 1950: THE BRIGGS CUNNINGHAM CAMPAIGN PHOTO ARCHIVE ISBN 1-882256-21-2
MARIO ANDRETTI WORLD CHAMPION DRIVER SERIES PHOTO ALBUM ISBN 1-58388-009-7
SEBRING 12-HOUR RACE 1970 PHOTO ARCHIVE ISBN 1-882256-20-4
VANDERBILT CUP RACE 1936 & 1937 PHOTO ARCHIVE ISBN 1-882256-66-2
WILLIAMS 1969-1998 30 YEARS OF GRAND PRIX RACING PHOTO ALBUM ISBN 1-58388-000-3

RAILWAYS

CHICAGO, ST. PAUL, MINNEAPOLIS & OMAHA RAILWAY 1880-1940 PHOTO ARCHIVE ISBN 1-882256-67-0
CHICAGO & NORTH WESTERN RAILWAY 1975-1995 PHOTO ARCHIVE ISBN 1-882256-76-X
GREAT NORTHERN RAILWAY 1945-1970 PHOTO ARCHIVE ISBN 1-882256-56-5
GREAT NORTHERN RAILWAY 1945-1970 VOL 2 PHOTO ARCHIVE ISBN 1-882256-79-4
MILWAUKEE ROAD 1850-1960 PHOTO ARCHIVE ISBN 1-882256-61-1
SHOW TRAINS OF THE 20TH CENTURY ISBN 1-58388-030-5
SOO LINE 1975-1992 PHOTO ARCHIVE ISBN 1-882256-68-9
TRAINS OF THE TWIN PORTS, DULUTH-SUPERIOR IN THE 1950s PHOTO ARCHIVE ISBN 1-58388-003-8
TRAINS OF THE CIRCUS 1872-1956 PHOTO ARCHIVE ISBN 1-58388-024-0
WISCONSIN CENTRAL LIMITED 1987-1996 PHOTO ARCHIVE ISBN 1-882256-75-1
WISCONSIN CENTRAL RAILWAY 1871-1909 PHOTO ARCHIVE ISBN 1-882256-78-6

TRUCKS

BEVERAGE TRUCKS 1910-1975 PHOTO ARCHIVE ISBN 1-882256-60-3
*BROCKWAY TRUCKS 1948-1961 PHOTO ARCHIVE** ISBN 1-882256-55-7
DODGE PICKUPS 1939-1978 PHOTO ALBUM ISBN 1-882256-82-4
DODGE POWER WAGONS 1940-1980 PHOTO ARCHIVE ISBN 1-882256-89-1
DODGE POWER WAGON PHOTO HISTORY ISBN 1-58388-019-4
DODGE TRUCKS 1929-1947 PHOTO ARCHIVE ISBN 1-882256-36-0
DODGE TRUCKS 1948-1960 PHOTO ARCHIVE ISBN 1-882256-37-9
JEEP 1941-2000 PHOTO ARCHIVE ISBN 1-58388-021-6
JEEP PROTOTYPES & CONCEPT VEHICLES PHOTO ARCHIVE ISBN 1-58388-033-X
LOGGING TRUCKS 1915-1970 PHOTO ARCHIVE ISBN 1-882256-59-X
*MACK MODEL AB PHOTO ARCHIVE** ISBN 1-882256-18-2
*MACK AP SUPER-DUTY TRUCKS 1926-1938 PHOTO ARCHIVE** ISBN 1-882256-54-9
*MACK MODEL B 1953-1966 VOL 1 PHOTO ARCHIVE** ISBN 1-882256-19-0
*MACK MODEL B 1953-1966 VOL 2 PHOTO ARCHIVE** ISBN 1-882256-34-4
*MACK EB-EC-ED-EE-EF-EG-DE 1936-1951 PHOTO ARCHIVE** ISBN 1-882256-29-8
*MACK EH-EJ-EM-EQ-ER-ES 1936-1950 PHOTO ARCHIVE** ISBN 1-882256-39-5
*MACK FC-FCSW-NW 1936-1947 PHOTO ARCHIVE** ISBN 1-882256-28-X
*MACK FG-FH-FJ-FK-FN-FP-FT-FW 1937-1950 PHOTO ARCHIVE** ISBN 1-882256-35-2
*MACK LF-LH-LJ-LM-LT 1940-1956 PHOTO ARCHIVE** ISBN 1-882256-38-7
*MACK TRUCKS PHOTO GALLERY** ISBN 1-882256-88-3
NEW CAR CARRIERS 1910-1998 PHOTO ALBUM ISBN 1-882256-98-0
PLYMOUTH COMMERCIAL VEHICLES PHOTO ARCHIVE ISBN 1-58388-004-6
STUDEBAKER TRUCKS 1927-1940 PHOTO ARCHIVE ISBN 1-882256-40-9
STUDEBAKER TRUCKS 1941-1964 PHOTO ARCHIVE ISBN 1-882256-41-7
WHITE TRUCKS 1900-1937 PHOTO ARCHIVE ISBN 1-882256-80-8

TRACTORS & CONSTRUCTION EQUIPMENT

CASE TRACTORS 1912-1959 PHOTO ARCHIVE ISBN 1-882256-32-8
CATERPILLAR PHOTO GALLERY ISBN 1-882256-70-0
CATERPILLAR POCKET GUIDE THE TRACK-TYPE TRACTORS 1925-1957 ISBN 1-58388-022-4
CATERPILLAR D-2 & R-2 PHOTO ARCHIVE ISBN 1-882256-99-9
CATERPILLAR D-8 1933-1974 INCLUDING DIESEL 75 & RD-8 PHOTO ARCHIVE ISBN 1-882256-96-4
CATERPILLAR MILITARY TRACTORS VOLUME 1 PHOTO ARCHIVE ISBN 1-882256-16-6
CATERPILLAR MILITARY TRACTORS VOLUME 2 PHOTO ARCHIVE ISBN 1-882256-17-4
CATERPILLAR SIXTY PHOTO ARCHIVE ISBN 1-882256-05-0
CATERPILLAR TEN INCLUDING 7C FIFTEEN & HIGH FIFTEEN PHOTO ARCHIVE ISBN 1-58388-011-9
CATERPILLAR THIRTY 2ND ED. INC. BEST THIRTY, 6G THIRTY & R-4 PHOTO ARCHIVE ISBN 1-58388-006-2
CLETRAC AND OLIVER CRAWLERS PHOTO ARCHIVE ISBN 1-882256-43-3
ERIE SHOVEL PHOTO ARCHIVE ISBN 1-882256-69-7
FARMALL CUB PHOTO ARCHIVE ISBN 1-882256-71-9
FARMALL F- SERIES PHOTO ARCHIVE ISBN 1-882256-02-6
FARMALL MODEL H PHOTO ARCHIVE ISBN 1-882256-03-4
FARMALL MODEL M PHOTO ARCHIVE ISBN 1-882256-15-8
FARMALL REGULAR PHOTO ARCHIVE ISBN 1-882256-14-X
FARMALL SUPER SERIES PHOTO ARCHIVE ISBN 1-882256-49-2
FORDSON 1917-1928 PHOTO ARCHIVE ISBN 1-882256-33-6
HART-PARR PHOTO ARCHIVE ISBN 1-882256-08-5
HOLT TRACTORS PHOTO ARCHIVE ISBN 1-882256-10-7
INTERNATIONAL TRACTRACTOR PHOTO ARCHIVE ISBN 1-882256-48-4
INTERNATIONAL TD CRAWLERS 1933-1962 PHOTO ARCHIVE ISBN 1-882256-72-7
JOHN DEERE MODEL A PHOTO ARCHIVE ISBN 1-882256-12-3
JOHN DEERE MODEL B PHOTO ARCHIVE ISBN 1-882256-01-8
JOHN DEERE MODEL D PHOTO ARCHIVE ISBN 1-882256-00-X
JOHN DEERE 30 SERIES PHOTO ARCHIVE ISBN 1-882256-13-1
MINNEAPOLIS-MOLINE U-SERIES PHOTO ARCHIVE ISBN 1-882256-07-7
OLIVER TRACTORS PHOTO ARCHIVE ISBN 1-882256-09-3
RUSSELL GRADERS PHOTO ARCHIVE ISBN 1-882256-11-5
TWIN CITY TRACTOR PHOTO ARCHIVE ISBN 1-882256-06-9

**This product is sold under license from Mack Trucks, Inc. Mack is a registered Trademark of Mack Trucks, Inc. All rights reserved.*

All Iconografix books are available from direct mail specialty book dealers and bookstores worldwide, or can be ordered from the publisher. For book trade and distribution information or to add your name to our mailing list and receive a **FREE CATALOG** contact:

Iconografix, PO Box 446, Hudson, Wisconsin, 54016 Telephone: (715) 381-9755, (800) 289-3504 (USA), Fax: (715) 381-9756

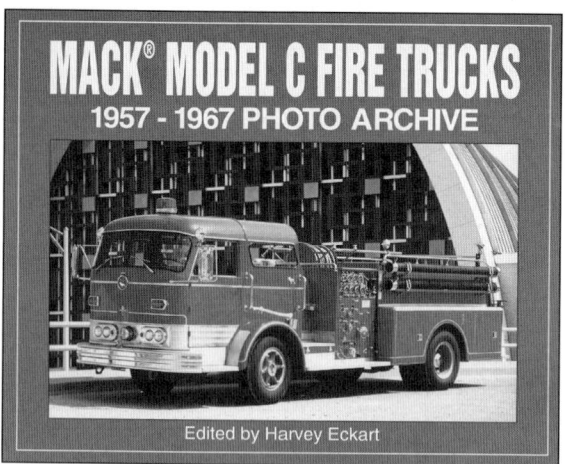
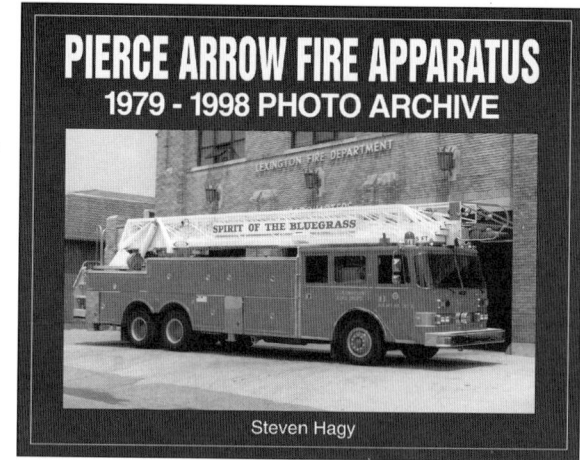

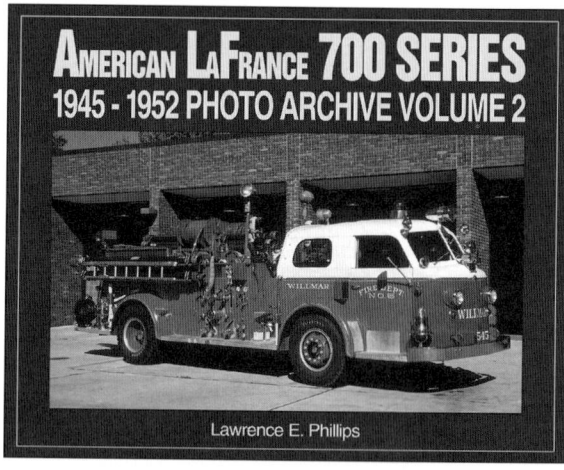
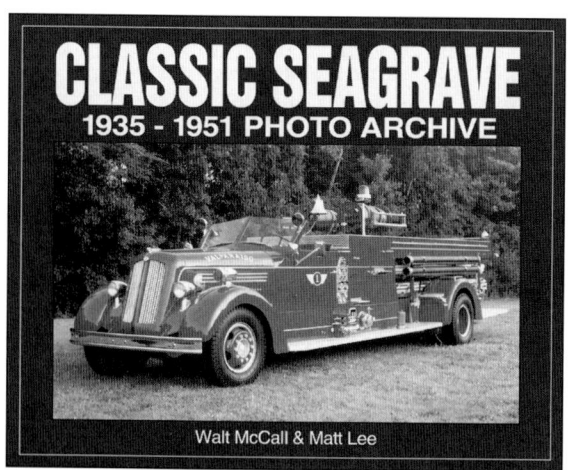
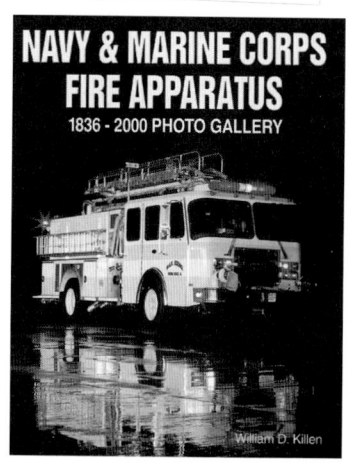
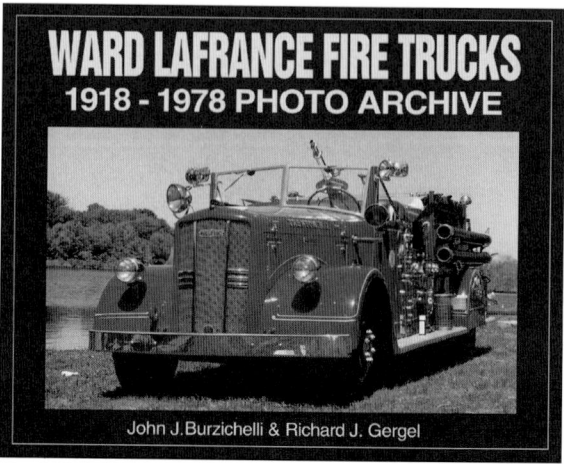